Forces of Nature

Forces of Nature

The Principles of Rotation

Tom Tong

Writers Club Press
San Jose New York Lincoln Shanghai

Writers Club Press
an imprint of iUniverse.com, Inc.

For information address:
iUniverse.com, Inc.
5220 S 16th, Ste. 200
Lincoln, NE 68512
www.iuniverse.com

ISBN: 0-595-19110-X

Printed in the United States of America

This book is dedicated to God and to

those whom have devoted their lifetime

in studying His work.

Contents

Preface

Long before writing this book, I had always been fascinated with things that were round. I could see the intricacies of roundness in the wheels of a car, or the way a bicycle balances in motion, or how a spinning top seems to defy gravity.

We might begin by asking why round forms are so common in nature? On a fundamental level, be it microscopic or macroscopic, the interplay between matter and motion is closely aligned to the concept of angular and circular behaviors. Such a common theme running through the universe raises some interesting questions and perhaps it's time to re-examine this universe from a new perspective.

Nicolaus Copernicus revolutionized the world view by observing that the Sun is sitting in the center of the solar system, with the other planets and earth orbiting around it. Rutherford envisioned a similar model for the submicroscopic structure of the atoms. The everyday world around us contains much that is round. We can see it in the ripples of the water, the spiral of the hurricane or even the gears and the pulleys of our simplest machines. It's a shape which possesses diverse capability across a wide range of phenomena.

Here is an example. Consider a two-dimensional circle. We may begin by rotating about its center, at angles of 11°, 33°, 55°, 90° or any other angles we may wish. It quickly becomes clear that the final appearance is the same as the original, no mater how much an angle the circle was rotated. Analogously, *the rotation of a circle through any angle is the same to any observer.* Take note that we are assuming that we are not interested in any particular point of the circle in our preliminary discussion, which will be highlighted with detail in the later part of this book. Interestingly, this *principle of rotation* intricately fits any kind of figure that is or close to something having a round figure. We may study the sphere as the next example and still be able to arrive at the same result. More interestingly for a sphere, the appearance remains the same even as we begin to change the angle of the axis of rotation. Such constancy in orientation actually allows the entire system to conserve all forces of nature. Having said this, it is not difficult to see that a round figure seems to be the perfect shape chosen for the physical laws of nature.

For the purpose of argument, we may test our analogy on other objects such as a cube or a pyramid for example. If we employ the same experiment that we have used for a sphere, we may not be able to get the same results for a cube, pyramid, square, triangle, polygon or any other shapes. Specifically, a change in the angle of the axis of rotation or the angle of orientation actually puts the object into a different appearance. A triangle may require 120° to restore its original appearance while a square will need 90°. It follows that an octagon with eight sides probably needs 45° to achieve the same appearance. Consequently, the more sides a figure has, the lesser the angle of restoration.

The dimension of a circle or a sphere is another interesting point to note. Needless to say, it is understood that the circumference of a circle given by $2\pi r$, area is πr^2 and the volume of a sphere is $4\pi r^3/3$ where r is the radius from the center of the figure. All these objects have something in common; their dimension is only dependent on one variable—the radius–and they obey the constant π. That is to say, the change in the size is directly proportional to the change in radius. Newton's discovery of the inverse-square law for the universal law of gravitation serves to be a good example of the dependence on the radius between two masses that describes the ellipses for the planetary systems.

In my earlier school days during a visit to a Science Center, I had my first glimpse at a model of our Solar System. The movements of those planets orbiting a central star were so wonderfully rhythmic, it was as though they were dancing to a beautiful song. Despite huge differences in size and composition, they were able to interact in endless harmony and peace. Given such beauty and tranquility in the realm of space, one cannot help but realize that the work done by the Creator must have been very exhaustive. It would be a shame for us not to appreciate such wonders of form and motion.

Gravity has long been known to control the motions of the heavenly bodies and is one of the fundamental forces in the Universe. That being the case, it is worth asking that most fundamental of questions: "What makes gravity?" Can gravity be "made" in the first place? We all know that we can produce a force by impressing on something. We can also feel force when it acts upon us. What then, produces gravity?

Centuries ago, Newton summed up gravity in his inverse-square law. This law easily explained that the force between two masses decreases by an inverse proportion by the square of their distance apart, and that the force of attraction between them could be derived mathematically. But while Newton's formula described this force, he did not explain it.

During the early part of the twentieth century, Einstein took this challenge and attempted, with a radically new approach, to explain gravity in his General Theory of Relativity. The intrinsic part of this theory was developed a decade after his famous Special Theory of Relativity. While trying to describe his a major component of General Relativity, the *Principle of Equivalence*, he exclaimed, "it was the happiest thought of my life!"

The essence of this principle is that the effects of gravity and acceleration are equivalent and indistinguishable. In other words, in Einstein's cosmic view the laws of physics are the same everywhere and at all times, despite any effects of motion and gravity. From this assumption, he deduced that the mass of an object could cause space to warp and light to bend. Therefore, anything that falls under the influence of gravity will follow a path defined by this principle such that we are unable to distinguish whether it is gravity or acceleration which acts on the falling object.

The development of these two laws has brought about a better understanding of the forces of nature. However, if we investigate further, we find that neither law explains the components of this force of nature. Newton's inverse-square law defines gravity, in a measurable sense, by way of a mathematical formula Einstein's view states the effects caused by this force of nature: that space warps in the presence of mass, while light bends as the proof of this theory. But neither of these theories adequately explains the components of this mysterious force known to us as gravity? Perhaps this is why that all forces of nature remain an enigma while scientists work restlessly to compile all known forces of nature into one single unified theory.

The attempt of this book is not to challenge prevailing theories but to describe the fundamentals for all the known forces of nature and, hopefully, to help the reader arrive at a better understanding and a more comprehensive picture of the world around us. Gravity is only one of the known forces of nature. The scope here is not just limited to gravity even though gravity will be emphasized in one of the chapters. I will also examine the logic behind all these mysterious forces by looking at the basic foundations for all these forces. Rather than using esoteric formulae and bombastic mathematical derivations, I have attempted to treat the subject in the most elementary manner.

Introduction

Forces of Nature

The laws of God are the same among themselves.

Modern theoretical physics has seen several revolutions. As in most areas of study, new ideas are founded out of the old. In physics—the study of the fundamental laws of nature—mankind attempts to describe the world that he sees around him. Scientists have worked exhaustively for centuries to harness the potential of these laws of nature in mechanisms together with the limited resources available to us. Armed with inquisitive minds, scientists have reached many milestones in understanding the world that lies around and beyond us. The immediate challenge for them now is to arrive with a single theory or a "theory of everything" which describes such a complex Universe. The search for such a theory has become a sort of Holy Grail for the scientific world. Many great thinkers like Isaac Newton, Albert Einstein, Stephen Hawking and others have devoted their efforts to the study of science. By achieving breakthroughs in understanding, they hoped their discoveries would bring deeper insights and, hopefully, better lives to their world of the time, and to future generations.

The birth of our Universe is believed to have begun with a Big Bang. This being so, skeptics will inevitably ask what was before this birth? Is our Universe finite or infinite? Where did all the huge amount of matter come from? The questions can be as endless as the anatomy of the Universe. It is these unanswered questions which keep scientists in an endless pursuit of an ultimate philosophy of science. If you look at the big picture, the study of science—be it physics, chemistry, cosmology or biology—you will never be able to dismiss the idea that there is a correlation between all the genres of science.

Physics is a dynamic subject. In fact, it is the study of physics which has made modern engineering possible. Like the study of any natural science, physics tend to uncover the secrets of the universe. But physics is also the fundamental building block behind the other natural sciences. To properly study chemistry, it is necessary to understand the fundamental structure of atoms and molecules, which therefore lean heavily on physics. Biology, too, relies on chemistry and physics to explain the intricate workings of living things. In this way, physics is vital because it provides the theoretical framework for the study of all natural sciences.

The objective of physics is to study the composition and interactions of matter. All matter in our Universe comes from pre-existing matter in that endless recycling dance we call the *law of conservation*. Here on Earth, you can see it in the cycle of living-and non-living things–from the life stages of animals and plants, to cellular processes like photosynthesis or the Krebs' cycle, to planet-wide transfers like the carbon or water cycles. In space, physicists can see the birth of one star and the collapse of another.

On Earth, life exists nearly everywhere. We need not restrict the word "life" to just living things. The trees and flowers have lives because they grow and bloom. The Sun has life because it spins and evolves with solar activity. Even a piece of metal has life because the atomic particles inside it are constantly in motion. If your eyes had the power of an "atomic magnifying glass", you could see all these moving particles, and the huge spectrum of light that is constantly bombarding the world around us, and even down to your very cells.

By this definition, every form of matter has a life of its own. The difference is that these various forms exist differently and therefore behave differently. But they do behave. Just as we have our own statutes and legislations to obey, so they have their own laws to obey too. Matter, be it solid, liquid or gas (or the lesser-known fourth state, plasma) may behave differently with respect to each other, but they do have their own physical laws of nature to comply with. The common ground they share is motion. Trees grow, birds fly, snakes crawl, fishes swim, celestial bodies circle in perpetual orbits, electrons spin, and so forth. They obey the rules of nature, and any world without laws to govern their movements would be in total chaos.

1. The Principles of Rotation

God had it all round.

There is a simple and yet powerful law observable in nature which governs the entire Universe. Remarkably, this law has been able to dictate all the natural laws of motion without flaw or variation (the slightest negligence on its part would, of course, render the entire Universe into chaos.) A full understanding of this principle has remained an enigma for centuries, and has challenged scientists to find the one law or unified theory which describes all systems in motion in the universe. From the movement of a particle, living or non-living things, to the perpetual motion of the heavenly bodies, all objects behave as though a set of rules were written specifically for them. Birds can fly because their wings were designed to lift them into the air, fishes can swim because they have fins to propel them through the water and so forth. But who is capable of engineering such intricate designs and still able to execute them in such a complex world?

There are so many aspects of this fundamental law of nature to investigate. First, let us look at the physical "life" of things. In order for anything to have "life", that thing needs to have "food" or a source of energy. Just as that thing needs energy, so it needs forces to keep it in motion so that it can find more "food" or energy sources. In order to have motion, variants such as distance, direction, time taken, linear or angular, relative or irrelative motion need to be considered. And finally, there needs to be enough space so as to facilitate movements of the thing. To facilitate movements within a certain space, there needs to be a geometry for the motions so that the thing can move for as long as is necessary. Whether these complex variables come from a Creator's hand or are an accident of formation, the marvel is no less great. Consider, for instance, if all of the astronomical bodies were to travel in a straight path. Imagine the extent of space that would be necessary in order to accommodate them? Would the states of matter be a constant if electrons are not held and made to orbit around the nucleus? Could "life" exist if the Earth were not rotating and orbiting around the Sun? It seems unlikely. Hence, the logical answer to these queries would be that everything would have to be cyclical and conform to the principles of rotation.

The principles of rotation allows everything to travel in a manner prescribed within a certain space without disrupting other nearby bodies in their original states of motion. In other words, a particle can be in constant motion without disturbing the rest of the variants. Electrons can therefore orbit an infinite number of times without altering the state of the matter they comprise. So does the motion of the planets, comets, stars go on without affecting the larger systems of which they are part. The periods of the motion can be a constant too. Best of all, all the effects of that motion becomes dimensionless and infinite. The distance covered would have been very far and the time taken would have been very long and the energy would not have been conserved if it had not been moving under the principle of rotation.

The principle of rotation also compels all other natural laws of forces to comply within the same principle. Consider again the planets acting under the gravitational influence of the Sun through a basic Law of Universal Gravitation. The distance traveled for a planet at a different point of the ellipse may be different but the time taken for its orbit is more or less a constant. Just trying to calculate the equivalent linear distance that the Earth has traveled since the birth of our solar system yields figures which are astronomical! Similarly, if you apply this idea to a simple pendulum, you will notice how energy is conserved in this *simple harmonic motion*. The principle of rotation, then, not only allows the motion of any thing to be conserved; it also allows natural laws of forces to be conserved as well.

Consider the example of a bicycle. Is the bicycle a very complicated machine? Superficially, it doesn't look to be. We can only see some gears, a chain drive, a set of pedals and, of course, a set of wheels. By just stepping on the pedals, we can force the gears to rotate one another with the help of the chain drive, and transform that human (mechanical) energy onto the wheels, finally moving the bicycle from one place to another without tipping over. Thought of in this way, it's an amazing process. Let us examine each component one by one. First, when we press the pedals, the gears are rotated in an angular motion. This angular motion produces a torque (rotating force) that drives the chain with a linear force. This force is then conveyed to the gear attached at the rear by transforming the force back to a torque to the rear wheel. (If you take a close look at the chain drive, you will also notice that the chain drive is made of numerous little cylinders that allow the chain drive to be bend.) When the rear wheel is being rotated, the friction between the tires and the ground will cause the torque to be transformed back to a force. Such is the power of the principle of rotation. As you can see from the above example, rotation and translation can function at the same time. The versatility of a round object in motion can really work wonders.

For a start, there are five states of rotation, which we'll be examining in this book:

1. The first state of rotation occurs when a rigid body undergoes rotational motion. In this case the tangential motion of each particle *increases proportionally* with the radius from the axis of rotation and unless the radius of the particle *in the axis of rotation* approaches zero, then the particle will also have an angular motion.

2. The second state of rotation describes a semi-rigid body undergoing rotational motion. In this case the tangential motion of each particle *increases proportion-ally* with the radius from the axis of rotation until a

certain particle and *decreases* after that particle if everything else is to remain the same. When an external force acts on the system and the internal force is sufficiently removed, the particle having the largest radius will move *away tangentially with respect to its system* and *an angular motion about its own center*. However, in the absence of external force, or when the net force is zero, then the body will act in the same manner described by the first state of rotation.

3. The third state of rotation describes a fluid body undergoing rotational motion. Here, the tangential motion of each particle is *inversely proportional* to the distance from the axis of rotation such that their laws of motion are definable within the theorem of the center of mass and the conservation law of angular momentum. However, in the absence of an internal force, the particle will move away tangentially with respect to its system with an angular motion about its own center.

4. The fourth state of rotation describes a two dimensional system which experiences rotational and translational motion *simultaneously* definable within the first three states. In this case, the particles in the center will translate with respect to a reference system and unless the radius of the particles in the center approaches zero, then they will have angular motion of its own. The other particles, having a certain distance included in the same space, will travel such that they exhibit the properties of a particle and a wave as they rotate and translate at the same time.

5. The fifth state of rotation describes a system which experiences rotational and translational motion simultaneously definable within the first three states. In this case, the particles will rotate along the plane of rotation and translate perpendicularly to it (for example, imagine a corkscrew). However, it is beyond the scope of this book to examine this state.

The above is a condensed description of the five states of rotation which will be addressed later in the book. Therefore, I urge the reader to skip this part and read on if you find it incomprehensible now.

2 Motion

The natural laws of motion continue to have the same direction if left to itself.

So what is motion? Motion can be described as a process of movement or the behavior of an object that has traveled over a distance (displacement) with time, as measured by an observer in a particular frame of reference. Only relative motion can be measured. Scientists use the term motion because it is generic and is simpler to visualize. By motion they are not particular about its direction or its magnitude. On the contrary, they are concerned with the results of that particular motion on the mass.

Motion can bring about a change in the state of matter. When an object is at rest, then it has only mass, but when it is moving we begin to describe its dynamics with terms like momentum, force, kinetic energy, or torque. All these are merely terms that scientists use to describe the magnitude and effect of the motion. As the speed gets higher, the energy gets larger such that it will "disturb" the state of the object. For instance, if we treat water with the right amount of heat, the molecules in the water will gain enough energy to escape in the form of steam or vapor. The motion of solids is as rigid as their shapes, either linear, rotational, or both. But a liquid can have turbulence, whirlpools, steam, waves, ripples, and so on. All these are driven by motion and the direction of the force.

Let us examine the concept of "force" again. Is force real or imaginary? Can we see force? Physicists always talk about force, but how clearly does the general reader understand the concept? If you hit a wall with your hand, you will feel the force because it hurts you. If you stand upright for too long, you will feel tired because gravity exerts a continual force on you. These types of forces come in pairs; an action and reaction that is stated by Newton's third law. Physicists try to measure these forces by various methods and attempt to quantify them into various categories. There are four fundamental forces that govern the dynamics of our universe, namely: electromagnetic force, strong nuclear force, weak nuclear force and gravitational force. All these are different in terms of magnitude, range of influence, positive or negative charges and so forth. Nonetheless, they have something in common and that is motion. Motion, it seems, is the basis of all force.

A force can play a different role when acting on different bodies. When multiple forces act on a single body, the resultant force will determine the path of that body's motion. Force can also cause a change in the state of motion and if there is no force acting on a body there will be no change in its speed. The difference between the terms 'speed' and 'velocity' is that velocity is *direction dependent*. The velocity of an object can change for two reasons; namely the change in speed over time or the change in direction. When the direction of a body's motion changes, its velocity changes, and hence the body is said to be accelerating or decelerating. Different shapes exhibit different behaviors when one or more forces is applied. Generally speaking, bodies in motion can be described as linear and rotational or both. This can be quite complicated especially in the field of space and time. Time is no longer an absolute value when calculated in space, but treated as a new co-ordinate in a four dimensions system. In Einsteinian mechanics, time becomes a value for an event—an event that is not only dependent on distance, but also on the propagation of light that concludes the event to an observer regardless of the frame of reference. This may seem a rather abstract concept, and to this day lots of scientists have spent vast amounts of time and energy studying the subject. However, it is not our concern in this book to explore any of the mechanics laid down by Einstein in his Special Theory of Relativity or General Theory of Relativity. Nonetheless, a good understanding of the Principle of Relativity will be helpful in the following topics.

Let us examine an object of mass m at rest in space; motionless and weightless. We say it has an inertia of mass m. Now we introduce a little force on it, just enough for it to gain a little velocity; then we have what we call momentum. With this simple translation in motion, we have seen that an object with an initial velocity 0 m/s of mass m turns into motion with a velocity v and this is formulated as mass *times* velocity which is momentum! If we continue with this process such that the velocity changes with time (in other words, acceleration) then the object will reach the next level known as force. While the acceleration remains constant, the velocity continues to increase. At this stage, we may discover that the object possesses kinetic energy. If the object is able to achieve to the speed of light c, then—voila!—the total mass of that object produces an enormous amount of energy such that we see the making of $E=mc^2$. While this example is merely a loose analogy, the objective here is to illustrate the interplay between mass and energy.

3 Paradox of a Merry-go-round

God never plays dice.
—A. Einstein

My fondest memories of childhood involve a playground, one of my favorites, being a merry-go-round. Back then, of course, I never imagined that the simple rotation of this carousel actually could encompass such complex dynamics. The feeling of being rotated was indescribably exhilarating and sitting there during its motion inspired me with forming scientific thoughts.

On one particular day my childhood friend, Tom, and I were sitting together on the merry-go-round. I was on the outside, further away from the fixed axis, and he was sitting somewhere in the middle (nearer to the center). Oblivious to the surroundings, I could see him traveling alongside me. Relative to each other, there was no way we could have told that we were moving had it not been for the trees around us. With my eyes shut, I could not tell whether I was in motion except for the feeling of being thrown outwards (centrifugal force) while I tightly held on to the bar (centripetal force) so as to keep myself in same rotational motion (frame of reference). And yet, there was a force that propelled me to move forward and faster than Tom. Hang on a moment, did I say "forward and faster than Tom"? Was I really traveling faster? When I measured the distance that I have covered and the distance that Tom had traveled, it became apparent that I had to be traveling faster in order to arrive at the same point within the same time! How could this have happened when I did not make any extra effort to speed up (accelerate)? Then again, maybe Tom was traveling slower than me. Both ideas sounded logical—so which was true? By this time, I was really confused. One thing is certain, as Einstein had said, "God never plays dice".

The issue I wish to address here is not just a matter of who was traveling faster or slower. The problem that attracted my attention was the complications that arise from the situation. If someone—within the same frame of reference—was to travel faster or slower than me, then it should be easily understood that the kinetic energy is different for each body within that same rotational reference frame about a fixed axis. Since rotary motions exist almost everywhere, the significance of a particle in relative rotary motion cannot be treated like the simpler examples of a particle in linear motion.

Take for example two men sitting inside a train traveling in a uniform motion. Everything that happens on that train can be treated in the same reference frame. Suppose that one of them starts to take a stride down the train, an external force of the man's walking legs causes a change in speed. Hence, the second man is seen to be accelerating from the first.

Now let's return to the merry-go-round again. Tom was closer to the center and I was on the circumference. When we looked at each other, both of us appeared to the other to be traveling at the same speed. However, to an observer out-side the carousel, I would have appeared to be traveling faster. In other words, *things that appear to be the same from the same rotary space are not the same when it comes to rotary motion outside that space.* Furthermore, the fact that my greater speed did not require an additional force to be applied on me makes the dynamics behind the merry-go-round even more interesting.

We may also examine at the carousel from a different per-spective which gives us even more peculiar results. Suppose we ask Tom (center) to sit in the center and face north before the start of the rotation. In this case, we need to con-sider his orientation during motion, with the reader deemed to be an *observer* on the ground. When the merry-go-round starts to rotate (let's assume counter-clockwise throughout), say 90^0, you will note the change in Tom's orientation. By this time, he is facing westward. If we continue the rest of the process at intervals of 90^0, the directions he faces will follow south, east and finally north respectively. Although

Tom (center) sits in the center, his orientation continues to change with the angle. The analogy here is simple. *Unless the radius at the center reaches zero, any point other than zero will have an angular motion.*

Suppose I stand on the circumference of the carousel facing towards the center-southward (also facing Tom) and having a certain position on the y-axis. When the carousel is being rotated in a counter-clockwise motion, I continue to face the center throughout the rotary motion. From my perspective, I could say that the Tom, who is sitting at the center of the carousel will appear to be rotating in a counter-clockwise direction relative to me. Here's why. At every interval of 90^0, my direction continues to change from south to east to north to west and finally back to south again. Strictly speaking, while having traveled a certain distance from the center, my direction changes proportionally when my position changes with respect to the center. Simply speaking, from the reader's (external) point of view, I should be "rolling" about the center of the carousel.

Hence, *in the absence of an internal force, a point having a certain radius from the center will have a tangential velocity about the center of the system and at the same time an angular velocity about its own center in the same direction of the rotation.* When the radius of the point *itself* reaches zero, then we can conclude that the point will only have a tangential velocity about the center of its system.

The above argument may seem odd and contrary to our common intuition of everyday experience. Let's examine the kinematics from another perspective. Suppose again I am still standing on the same position on the y-axis and facing south, but this time I will continue to face southward throughout the rotary motion. In other words, when the carousel is rotated, my position will change but my orientation remains fixed towards south. In order for me to do that, I *must rotate myself in a clockwise motion* so that for every angle that the carousel rotates, my facing remains southward. From the reader's (external) perspective, I must have a *clockwise spin of my own* to compensate for the counterclockwise motion of the carousel. In such a case, my angular kinetic energy about my own axis and tangential kinetic energy with respect to center of the carousel will be in opposite directions. Essentially, this will of course *oppose* the natural laws of rotations.

Specifically, unless influenced by an external force, the natural law of rotation of a point will have the same direction as the natural law of rotations of the system.

It might be worthwhile doing this simple experiment to illustrate the above analogy. Suppose we use a string tied to a ball and swing it horizontally around our hand. After that, we may release it and observe the motion of the ball. Common experience tells us that the ball will fly tangentially at the point of release, which is true. However we are not able to observe whether the ball actually spins or not as the ball gets further away from us. In such a case, we should swing in a vertical manner and release it while it's going up so that gravity will act on it and the ball will not fly as far as it could. Upon release, you might be surprised to see that the ball not only flies tangentially but actually spins and *the direction of the spin is exactly the same direction of the swing*!

In an earlier chapter, we discussed the principle of rotation. The most important part of this principle to be noted here is that particles in the same rotating reference experience different velocity and therefore, different kinetic energy. These differences cannot just be fixed simply by a rule of thumb. The total rotating energy is derived by the sum of the particles acting in that system. However, this differential in the behavior of particles can bring about a change of the state of the matter of which they are in. Hence, different types of matter behave differently when acting in a rotational motion. We will examine this more specifically under the four states of rotation motion.

4 Frame of Reference

Observations are meaningless without a theory to interpret them.
—Lyttleton, R.A.

In order to describe a body that is in motion, a car for example, we need to establish a co-ordinate frame called the frame of reference. It allows us to make measurements relative to the original reference frame. Otherwise, measurements taken from different observers could appear to be totally different, especially if their reference frames accelerates relative to one another. This frame of reference is particularly useful in the Principle of Relativity. It is important because we need to establish a point of origin and then another position after traveling in a positive direction along the same line. Only then can measurements be made and conclusions derived. Without a reference frame, descriptions of motion become meaningless.

An object is said to be in motion relative to another when its position, measured relative to a second body, is changing with time. Alternatively, if this relative position does not change with time, the object is said to be at rest. Motion and rest are relative concepts; they depend on the condition of the object relative to the body that serves as a reference. Therefore, the state of the reference body can affect the results of the measurement. This book and you are *at rest* relative to your house but are moving relative to an observer on the Moon or the Sun because of the Earth's orbit. When a car passes your house, you can say that the car is in motion relative to your house. On the other hand, the driver in the car might say that your house is moving relative to his car, but in an opposite direction.

Let's look at a simple example. Tom is driving a car with a speed of 50 km/h. We'll suppose that he has a vintage car which lacks a speedometer so he doesn't know the actual speed at which he is traveling. The only way to do this is to measure the things that are moving around him. If he observes only the objects with him *in* the car, he will conclude that everything is traveling at a speed of 0 km/h. This is because the things around him are in the same frame of reference. Using only these objects, he may conclude that he is stationary. But if he picks an object outside his vehicle, with a little measurement and simple arithmetic he will be able to calculate that he is traveling at the speed of 50 km per hour. Now imagine Paul standing at the side of road (inertial reference frame) and seeing Tom drive past. Assuming that he has a speed meter, he reads it and measures that Tom is driving at the speed of 50 km/h. Now if Paul wishes to catch up with Tom (not knowing the Principle of Relativity) and rides on a motorcycle also moving at 50 km/h, he will find that he will never be able to catch up with Tom (since Tom has already moved at a certain distance)!

5 Principle of Relativity

When you are courting a nice girl an hour seems like a second. When you sit on a red-hot cylinder a second seems like an hour. That's relativity.
—Albert Einstein

So what is the Principle of Relativity all about? Newton stated, in formulations of his laws of motions, *that the motions of bodies included in a given space are the same among themselves, whether that is at rest or moves uniformly forward in a straight line.* Specifically, this means that if an experiment is performed in a train for example, all observations in the train will appear the same as if the train is not moving, provided the train is moving uniformly in a straight line.

In an analogous statement, Einstein defined the Principle of Relativity as *a mass m moving uniformly in a straight line with the respect to a co-ordinate system K, then it will also be moving uniformly and in a straight line relative to a second co-ordinate system K', provided that latter is executing a uniform translatory motion with respect to K.* In retrospect, an assumption that exists in a reference-body K, whose condition of motion is such that the Galilean law holds with respect to a particle left to itself and sufficiently far removed from all other particles moves uniformly in a straight line. With reference to K the laws of nature were to be as simple as possible. But in addition to K, all bodies of reference K' should be given in this sense, and they should be exactly equivalent to K for the formulation of natural laws, provided that they are in a state of *uniform rectilinear and non-rotary motion* with respect to K. In another words, this is only valid in a non-accelerating reference frame or otherwise known as inertial frame of reference. The above is an abstract from the book *Relativity* by Albert Einstein.

In order to attain the greatest possible clarity, it is important to have full understanding of what we mean by uniform (non-accelerated) motion in a straight (linear) line. Imagine that you are on a plane that travels at a speed of 1,000 km an hour. It may sound fast by our standards, but to an observer onboard, everything would appear as normal. Your cup of coffee and your newspaper would lie motionless beside you. There would be no way to tell whether the plane is in flight or not. You could drop an apple, walk around or do any kind of experiments that you wish and still not able to tell whether your plane is in flight or parked. This is the essence of *uniform motion in a straight line.* As long as the plane does not roll or change its speed or direction, the laws of physics will remain the same in a given space. This is the significance of the *principle of relativity.*

Now let us try to understand what Galilean law (Galilean principle of relativity) means. Let's resume our flight and do some "thought" experiments on the plane. Suppose that you throw an apple towards the front at a speed of 100 km an hour, common sense and our intuition tell us that the apple should travel at the speed of 100 km an hour relative to you and 1100 km an hour (1000+100) if I were to measure it from the ground. In both cases, they are all true because if I am on the ground and measure the speed of the apple, the speed has to be 1000+100 km an hour in order for the apple to travel faster and reach the cockpit of the plane. Relative to you, the apple has to travel an additional of 100 km an hour so as to reach the front. Hence, both analogies are synonymous. This is known as the *theorem of addition of velocities* under Galilean principle of relativity.

Next, we shall attempt to understand Einstein's *special relativity* and see how special it really is. In the *special theory of relativity,* there are two underlying principles; namely

1. The laws of physics are the same for all observers traveling uniformly in a straight line (this is the same as Galilean relativity).

2. The speed of light *c* in vacuum is the same for all observers traveling uniformly in a straight line (this is the "special" aspect).

Since we have understood the first principle, I will go on by explaining the second postulate. Suppose again that you are on a jet plane this time and travel at one third the speed of light, 100,000 km per second. If I shine my flashlight in the same direction of your flight, the beam of light will travel at the speed of 300,000 km per second past you. Galileo predicts that the speed of light *relative to you* will be 300,000 km/s less 100,000 km/s=200,000 km/s. In another words, you will see the light traveling at a merely 200, 000 km/s or rather you have "slowed" down the speed of light!

Contrary to Einstein's belief, he noted that if the laws of physics are to be the same and the speed of light is to be a universal constant, the speed of light should be the same for all non-accelerated observers. Strictly speaking, the speed of light will remain the same regardless of the observer's motion. Therefore, you should see the speed c as the same speed that I have measured it from the ground. The idea seems odd but it turns out to be both profound and significant. However, there are two conditions that need to be satisfied in the theory of relativity; namely, it has to be ***non-accelerated observer (reference frame)*** moving in *a **straight line.***

6 Mass

The quantity of matter is the measure of the same, arising from its density and bulk conjointly.
—Newton

Newton's first law: Inertial mass is a property of a body that resists any change in its velocity.

The dynamics of an object can be quite complex if we do not idealize its mechanics in the form of particles. However even with such simplicity, when handling objects of different masses, shapes and motions, a simplified picture does not eliminate the complex dynamics of the total mass between different objects in motion. These complications are nothing more than the works of the most fundamental laws of nature.

The simplest form of a complicated object is probably a rigid body. It is an object in which the forces between the atoms are so strong and of such nature that its shape and volume stays the same in motion even when a force is implied on it.

The motion of a rigid body can be linear, rotational or both. Let us first look at a rigid body rotating about a *fixed axis*. If the object is a sphere, than we can simplify the sphere in the form of a rod and likewise, into the form of particles which can be treated as if they constitute the whole mass of the sphere. Further to this, we can idealize this particle as a point that moves in a constant radius from the fixed axis which is in a plane perpendicular to this axis, herein known as the *plane of rotation* or rotation in two dimensions.

When a sphere spins about a fixed axis, all particles having the same radius rotating perpendicular to that axis will produce a force equal in magnitude, such that a center or a mean position of the object is definable in the theorem of the center of mass. You may compare other particles in the same object having a different radius and still be able to arrive at the same position. However, this theorem does not limit or need to be a point in the material itself. In other words, if we are to consider two objects that are being held together, their mean position could be the average between the distances of their centers of mass.

What this means is that when many forces act on a mass of particles, the sum of all these forces on the total mass of the body results in a net force which produces acceleration at a point as if the whole mass were concentrated there in one place. This point is called the center of mass. In this context, the mass described has two different properties.

The first is a measure of a resist in change in velocity and the latter is an expression of an object's gravitational attraction to other objects.

7 Rotations

The natural laws of rotation have the same direction if left to itself.

An object is said to be in uniform circular motion when it travels in a circle at a constant speed. Although the speed is constant, the velocity continually changes as the direction changes. Hence, this motion is described as an accelerated motion and this acceleration tends to be directed toward the center. Why is this so when the centrifugal force seems to be directed outwards? This effect is caused by the change in direction of the tangential velocity of the body in motion. In order for the direction of a body's motion to be changed, an external or internal force must be applied._Hence, the greater the change of the angle, the greater centrifugal force will be experienced. On the contrary, if the object's final tangential velocity about a center is less than the initial tangential velocity, then the object will experience a deceleration.

Imagine being are in a car and driving at a constant speed. Both your body and the car are in motion together. If you step on the brake, you will experience a force that pushes you forward. Actually, it wasn't really a force since there was no acceleration. It was the natural state of your motion that causes you to move forward while the car is slowing down. Relatively speaking, the deceleration of the car causes an equal acceleration on you and therefore, you felt a force. Likewise, while turning the car around a bend, your natural state of motion is straight while the car is angular and therefore, centrifugal force is "created".

Relative Circular Motion (Restricted State)

An object that is in a circular motion experiences more complex dynamics than one in a linear motion. The contrast between circular motion and linear motion is that particles experience different types of forces and their forces are *radius-dependent* with respect to a fixed axis within that body. In other words, an external force acting in a direction other than the original motion will result in a change in force such that the force of any particle is different with respect to one another about a fixed axis within the same body.

An object experiences linear motion only if no external force acts upon it to shift it from its original direction. The earlier chapter on Newton's Law of Inertia also states that a body at rest will stay at rest. What happens, then, when an object starts to rotate? In order to understand this situation beyond the centrifugal force that we have already discussed, we need to examine some general features of this object.

Let's begin by looking at a car that is traveling in a straight line. When the car reaches a bend (left turn), the driver makes a judgment and turns the wheel so as to follow the bend. If he decides that he wants to move straight, then he may continue to move straight and drive away from the bend of the road, which essentially means that he breaks away from the winding of the road. This however, by the virtue of his original state of motion, does not require extra force—assuming that there is no centripetal force acting on the car. In fact, by turning, the car actually causes a change in velocity, which in turn causes an acceleration towards the center. This creates centripetal force and thus, centrifugal force. Centrifugal force is in this context the force to move in straight line. In another words, the car has nowhere else to go but to follow the path of the road, which is a bend. The analogy in this context is that the car *has an option of traveling forward or takes the bend.* So if the car chooses to take the straight road, the force will only be a linear component. But if it chooses the bend, then there will be extra components of the forces governing the motion. These competitive forces will decide the ultimate resultant force for that motion.

If we expand our example a little, let's assume that it is a long road and the driver has no idea that it is bending. The road will appear to the driver to be straight, and the driver continues to drive in the same manner as before. Because the bend of the curve is so minimal, he hardly feels any of the centripetal force that is acting on him. Together with the fact that the road appears to be straight, he will conclude that the road is straight. Little does he know that he was traveling *tangentially* on a huge circle! Simply speaking, he *is* traveling on a straight road at the point of the curve. This is somewhat similar to our daily experience with gravity and the rotational motion that we are in. We don't really "feel" it even we know it is there. While this notion may seem ambiguous at this stage, it explains the first component of a particle in circular motion which is going to be helpful when we go further.

Next, we look at particle (our car, again) in a restricted frame. Imagine the above scenario, but this time including a motorcycle in the inner lane traveling parallel to the car. As the car meets a left hand bend the motorcycle, in order to avoid a collision (***anti-collision***), will have no choice but to make a left turn along the bend as well. This is what I mean by the rotational motion of a particle in a restricted frame. In contrast to our earlier discussion about the car's options of going straight or making the curve, a particle that is traveling in a restricted frame is not only constrained by the direction—as in the case of the car and the motorcycle—but also by the velocity that it is traveling within that frame. Relatively speaking, in a rigid body—be it a rod or a sphere—the velocity (angular or tangential) of a particle cannot be faster than another particle that is further than the former from the center of rotation. This should be true provided that, other things being equal, no additional force is impressed on it.

Now let's consider what actually causes a car to turn. You can say it is the wheel of the vehicle that allows the car to turn, which is partly true. But the real kinematics behind that turn is that the front wheels of the car have actually decided to take different direction and a shorter path against the original path. By changing the angle of the wheel, the right side of the car will take a longer route and the left side will have shorter route to travel (assuming turning in a counter-clockwise direction). However, the acceleration is the same for both sides of the front wheels. In such a case, the wheels will not be able to achieve a position of equilibrium at this point of time. As mentioned earlier in our anti-collision analogy, this will bring about a change in velocity in both sets of wheels, which then causes the car to turn. If the turning is too acute, as in the case of a car traveling at high velocity, both set of wheels will not agree on the velocity that they are traveling and will cause the car to slide. Of course, in a real car, both set of the front wheels have to agree on the direction or else the car will be out of control, as different sides of the car will try to take a different route with a different velocity. Fortunately, modern roads are made to bend in a gradual manner so that the frictional force between the tires and the ground is able to counter this centrifugal force and maintain stability while the car is turning. For discussion purposes, if we suppose that only one wheel is made to turn, leaving the rest of the wheels going forward, what will be the resultant motion of the car then?

There will be a war among the wheels such that the net force becomes the resultant force.

8 The First State of Rotation

Rigid Body

The first state of rotation occurs when a rigid body rotates, the tangential velocity of each point increases proportionally with the radius and unless the radius of the point in the center reaches zero, then the point should have an angular velocity itself.

Let us consider a rigid object, say a piece of metal rod, rotating through an angle θ about a pivot 0. Assume also that this rod is made up of many particles. However, for simplicity sake, we will just take two particles on the rod and examine their kinematics in circular motion.

Let two particles, 1 and 2 sit on the rod, each having a certain distance from the axis of rotation while 1 being in the middle and 2 being the furthest. If we measure the distances that they have covered after rotating through an angle and after a certain time, we will discover that both particles actually traveled a different distance relative to one another while the period remains the same. In fact, all the particles on this rod, which have a different radius from the axis of rotation will have covered different distances and, therefore, have different velocities. As such, the difference in velocities in turn causes different accelerations. For this reason, it is not easy to fix an inertial frame of reference at any particular point other than the one that is on the angular co-ordinate—the axis of rotation.

In an earlier chapter, we mentioned about the anti-collision of particles moving in a rotational motion in a restricted frame. By the behavior of such particles, let us describe this phenomena as *position equilibrium* between that of the particles. The particles traveling in a restricted state are therefore not allowed to move any faster or slower than the speed that they are formulated to travel with respect to the other.

If we assume that particle 2 which is on the outermost of the rod to have the maximum tangential velocity, then we can deduce that particle 1 can only travel slower than 2. Theoretically speaking, as we get nearer to the center, we can determine that the particle at the center of rotation will have a angular velocity close to 0 m/s. On the contrary, the said result is an illustration of the amplification of the magnitude of a particle in circular motion. If we assume that the speed of light c is the ultimate speed that any particle in the Universe can travel, particle 2 will then achieve the speed of light so that particle 1 will obey the law and travel slower with respect to 2. There are two catches here. First, what if particle 1 is already moving at the speed of light, what will happen to 2 then? Analogously, if we impress a force such that particle 1 moves at the speed of light, what will be the speed of particle 2 then? Will it still continue to move at the same speed or will it move faster or slower? Second, the context here is a restricted state of a rigid body. The dynamism will change once the system changes to a different state. This idea will be discussed in the following chapters.

Now that we have a little understanding of the first state of rotational motion, let's resolve one more issue before going any further. It is in the reader's best interest to have a comprehensive understanding about the relative motion of particles in an isolated system, regardless of their state of motion. If you put two particles in an isolated system and scrutinize them thoroughly, you will then be able to deduce and conclude the validity of the event through observations and experiments. One caveat: if you were able to treat the particles in an isolated system the same as the observer's reference frame, then the results would be dependent on the reference frame of the observer, which might invalidate them. The debate here is that if you "look" at the particles that are in motion from a different perspective, you will be able to understand the kinematics behind it. Your observations will be valid because *you* (your reference frame) will not be affected (in other words, will be independent) by the laws that are governing the observed reference frame for rotary motions.

One classic example is how the Earth was once treated as the center of the solar system. This tradition of thought goes back to the Greeks. During that time, Aristotle decreed that the logical position of the Earth was in the center of the Universe. It is quite easy to understand why such a theorem was accepted at that period of history. Observations made were synonymous with the orbits of the Sun and Moon. But when early astronomers began to study the orbits of the rest of the planets, they behaved in a very peculiar way. Some of the planets were observed to slow down, stop, move backward and then forward. Scientists couldn't explain this strange phenomenon but were unwilling to accept that there was a flaw in this theorem. In 1543, Nicolaus Copernicus published a book entitled, *On the Revolution of the Heavenly Spheres,* which shook the world. His new idea actually described the Sun as being the center of the solar system, similar to the model we have today. The moral of this story tells us that observations made in one reference frame have to be synonymous with another independent reference frame. It is good to study the behavior of a particular object in motion by having a reference frame fixed within that object. However, experiments conducted in an enclosed space have to be synonymous with an observer's reference frame outside that space so that the laws of physics will always be the same.

Here is another example. If two men are sitting in a car that is traveling at a uniform speed, both will see that everything is moving relative to them so their net speed (together with the things inside the car) is zero (*observer: they are moving at a uniform speed*). Let's assume that this is a vintage car with no speedometer and the windows are all closed. The two men have no way of telling whether they are moving or not. If they drop an apple to the floor of the car, that apple will still drop perpendicularly. It is only from the point of view of an observer outside the car, who knows that they are indeed moving at uniform speed. Now let us imagine that the car begin to accelerate, the relative speed is still the same between the two travelers and it is still the observer who can still tell the difference by the difference of the initial speed and the final speed.

The results taken from two different reference frames may yield somewhat different values but the derivations can still be the same. I will try to explain again. In an accelerated reference frame, the travelers *may observe* the same measurements as if the measurements were taken from an inertial reference frame. In other words, the travelers *may* not be able to tell whether their car is accelerating or not. It is the same as if the car was still moving uniformly (*observer: they are accelerating*). But there is a catch in this case. If one of them were to drop an apple, then they would notice a change from the earlier case. This time, because the car is accelerating, the apple will drop slightly behind them. On the contrary, if the car is decelerating, a dropped apple would fall slightly in front of them. It was as though some mysterious force had been impressed on the apple that causes it to move in front or behind (*observer: this effect is caused by the change in speed of the car while the speed of the apple remains the same at the point of release*). The two men start arguing. One claims that we are moving faster than the apple and so the apple fell behind us and vice versa. The other said that there was an external force on the apple that caused the apple to drop in front or behind (depending on the direction of the force) and when there is no force (other than the gravitational pull) then it will drop perpendicularly. Then one suggested maybe it is because the car is no longer in the moving in a straight line, it could be that the car is flying at angle such that the apple drops in a trajectory in the opposite direction of the motion. They are really confused this time, and so probably is the reader!

Let's make this situation even more interesting (or perhaps more confusing). Suppose this car is in space now and it doesn't have seats, so that the two travelers will be both weightless and motionless. They both conclude that there is a fictitious force that keeps them suspended in mid-air (*observer: lack of gravitational force*). Suddenly, the car moves forward and the travelers move backwards (equal and opposite) and they think it is an act of gravitational force (*observe: the car moves forward*). Then the car stops abruptly and the travelers move forward this time and hit the other side of the wall and they think it is the act of another external force (*observer: car stopped but the two men continued in forward motion)*. Then the car begins to rotate with the pivot at the other end of the car so that the travelers still experience a force acting on them on the same side of the car *(observer: act of centrifugal force)*. Can they tell the difference between the first situation and the latter? Imagine that the car stops rotating but this time, rotates with the travelers as the pivot. Can they tell the difference? They might, by the motion of the side of the wall. But what if the car is in the shape of a sphere? Could they still tell the difference then? Clearly, they wouldn't be able to. They would not even notice if the car (sphere) turned upside down! In this case, how could they explain all the fictitious forces that are happening? It wouldn't be possible. Only the observer (and you) can see that the "fictitious forces" are the acts of the motion behind it.

9 The Second State of Rotation

Semi-rigid Body

The second state of rotation occurs when a semi-rigid body rotates, while the tangential velocity of a point increases proportionally with the distance from the center, it decreases after that point, unless additional forces act on it.

Our next task is to explore the rotational effects of the particles in a semi-rigid body. Consider a string of negligible mass that is made up of beads, each bead representing a particle in a rotational system. Each point on the string is represented by particle 0, 1, 2 and 3 where particle 0 is at the axis of rotation and the rest having a certain distance away from particle 0 with particle 3 being the furthest. A force is applied to rotate the string, constant enough to ensure that all particles can accelerate to their desired position at any point of time.

Two scenarios can happen in such a case. First, if we increase the speed of rotation without increasing the length of the string, the beads at 3 will move so fast that at a certain point they will break away altogether, leaving the beads at 2 and 1 in the string. When that happens, the law of conservation of angular momentum dictates that with the shortened string of beads (and their reduced mass) the beads at position 2 will now inherit the momentum that 3 used to have, and hence 2 will experience an increase in momentum with the same final breakaway result, and so on down the rest of the particles.

This is a classic example of a breakaway when the centripetal force is being overpowered by the centrifugal force. In another words, the acceleration of particles on the extremity accumulates enough force to break away.

In the second scenario, let us assume that the force applied is at a constant. This time, instead of varying the force, we vary the length of string by shortening it. The shortened string means that there are fewer beads (particles) in the body and since angular momentum p is a product of mass m, velocity v and radius r, it means that the tangential velocity of the particles will increase, since the total angular momentum remains the same. The same result would have happened in the above-mentioned case if we continue to decrease the length (mass).

Similarly, if we were to increase the length (mass) of the body, angular momentum being the same, the extreme particles will experience a negative change in tangential velocity. In fact, this will cause 3 to decelerate and loses its *position equilibrium*. The result will be a shift of momentum from **3** to **2** (being maximum) in the body. This analogy tells us that the increase in mass in a rotating system will bring about a decrease in the total angular momentum of a rotating body and therefore, particles on the extremity will decelerate (lag) in the same system without a change in the force applied.

The above also tells us that in order to achieve *position equilibrium*, a certain torque needs to be maintained constantly. The presence of force also tells us that there is acceleration. Once beyond the *line of break*, the particle **3** will 'feel heavy' even without losing momentum. Similarly, if **3** were to move to **2** by way of a shortened string or relocation, it will feel lighter and move faster even if everything else has been unchanged. This actually happens with the help of the law of conservation of angular momentum. Thus, a shift in position is enough to create a change in energy and therefore, mass. However, in the above scenario, **2** is actually the fastest as compared to **1** and **3**. Can **1** be the moving the same as **3**, assuming they are equidistant from **2**? Does it also mean that if a body is being rotated at different speed, the different parts of the mass in that body will actually experience different tangential momentum?

The Lighthouse Experiment

Now let's perform another thought experiment to illustrate this point, which I'll call the "lighthouse experiment".

Imagine a lighthouse at a certain place on earth, and two observers—one located on the earth (the "earthman" we'll call him) and one located in space ("the spaceman"). We will also assume that the earth is an inertial reference system, in other words we ignore the effects of its spin.

As readers will know, a lighthouse works by rotating a beam of light at regular time intervals. For each second of a time, the light beam emitted by the source will make a revolution seen by our space-based observer at the same time as our earth-based observer. As the spaceman travels further out, he will reach a point where the distance between pulses exceeds a second, signifying that his point in space is more than 3×10^8 meters (because it takes one second to reach 3×10^8 meters assuming also that light does not have to travel in a trajectory under the influence of the gravity).

When the spaceman has reached this distance from the source, let's assume he stays in orbit so that he receives the same signal at equal time intervals of one revolution per second. The purpose of this is to make sure that the time on his watch is synchronized with the time on earth. Even if the light beam from the lighthouse starts to make more revolutions per second, the spaceman and the earthman will still agree on their time given by the signal but not the time on their watch.

To make things a little bit more interesting, let us move the spaceman another 3×10^8 meters further out, such that he will receive the signal one second late for every flash that he sees. This means that it will take two seconds on his watch before he receives the signal (two seconds per signal). Even though there is a one second difference (between watch and signal), he can still calculate the time on his watch to be one second faster than the signal.

At the same time, the earthman will agree with the time that is on the watch and the signal. Can the spaceman simply adjust his watch by one second later? Unfortunately not. If he does that, every signal he receives will still be one second late. Why? Because the watch he is using is independent of the light signal that is traveling towards him. In other words, his time in space is still the same time as that on earth (ignoring other influences). The reason why his watch does not synchronize with the time given by the light signal is because the *light-time* works on the different formula; distance/ speed of light *c*, while the watch works on a different mechanism. That is to say, the longer the distance gets, the longer the *light-time* it will be. The time-reference frame for the watch is therefore independent of the light-time.

Likewise, different watches behave differently, especially in different frames of reference. Even if you choose to use two identical watches, running on the same mechanisms, the time will still be different if one watch is made to travel and the other is left stationary. The reason being that the time which registers on the moving watch is still affected by the motion of the accelerating frame of reference.

To reinforce this idea, we need to remember that time is itself elastic. The time that we use to measure events that happen in our daily lives is a value that was invented centuries ago, since the beginning of civilization. We use time in the popular sense to meet people, meet schedules, to gauge our age and so on. Scientists also use conventional time to record the change of events with either respect to distance or other values versus time.

This value or form of measurement was determined with the help of the constant orbit of the Earth around the Sun. Simply speaking, if we were to live on Venus or Mars and use the local orbital route around the sun, then citizens of each planet would have a different time all their own, claiming that there is time lag with respect to others. A year on Venus would last 225 Earth-days, for instance, while the Martian year would seem interminable at 687 Earth-days. Hence, your age will be younger if you were to live on Mars.

Time Reference Frame

So what is "actual time"? Why is it a necessary quantity and how do we define it? It is all too easy to take such a measure for granted, since we employ it everywhere we go and for most things we do. But time as we know is only valid within a specific frame of reference on Earth. Consider, for instance, travel to another country. We have to adjust our watches–not because the duration of time is different (an hour in New York lasts as long as an hour in Singapore) but because the calibration of time in each place is different. The earth's spherical shape and rotational path means that high noon on one side of the planet is midnight on the opposite side, and the time zones in between are calibrated to suit the needs of the regions they cross.

So how can you apply a value, which is not a constant on a universal scale? The law of time that mankind uses simply does not comply with the time that *laws of nature, which* is why it doesn't work on God's physical laws of motion. Even if you were to use the time that is defined by the Special Theory of Relativity; which is only valid when time measured with respect to the speed of light; is not so applicable when light (supposedly to be a constant) travels at a different rate in a different medium not to mention, in a different geometry. It is just like asking you to measure a curve with a straight ruler. No matter how hard you try, there's a limit to the precision of your measurements. Now that we have identified the problem, we need to seek an answer. Is there a universal form of time; a value that you can use to record the change (or lack thereof) of events versus another value, and is invariant with respect to the frame of reference (rotary or linear, uniform or non-uniform) universally? This is a sticky question!

One of my science teachers once taught me a golden rule, Rule No. 1 being that if you want to solve a complex puzzle; the trick is to make it simple, or at least, try to think of it in the simplest form. Let me give you an example. If you want to determine the speed of a car, ask yourself if you would start by looking at the engine of the car, examining its gears, counting the pistons, valves and spark plugs, then examine the combustion process from the chemical energy to mechanical energy released and so on*** or would you just look at the horsepower of the engine and just deduce the highest possible speed that it can go.

Let's do the exercise together. First, we know that time is a variable. In order to treat a variable as an absolute, we must be able to arrive at a deduction where time becomes a constant, such that it is an invariant irrespective to the frames of reference for any observer. When that happens, the value of time becomes independent of the speed, direction, motion (rotary, linear or both) and does not get influenced these factors. By influence, I refer to external forces that are defined by the laws of physics.

We may therefore begin by treating this *time* in its simplest form and this *time* has a value of t (ignoring any units for the time being) and we shall call this universal time, t_u. Now that we have a universal time t_u, specifically it means that our new *time* that we use is the same wherever we go, be it on Mars or Jupiter (it is independent of the orbit of the planets). There again, we mustn't forget that the "formula" for this *time* can even work on the atomic level. If such being the case, then it will be easier for us to focus our attention in the area of our search. It should be in some form of index, coefficient or a constant of any kind that governs the *periodic motion for the fundamental laws of nature* and it is only valid within the same reference frame.

Before I confuse the reader any further, let me reinforce the statement again. *The universal time t_u, is independent of any type of motion and is the same to all observers regardless of their reference frames.* For that reason, the motion becomes the function of time. It is not the time that we use in our daily lives, so you won't have to change your watch and clocks. The time that we use is only good when you use it on Earth (having the same frame reference, direction and the same motion). This kind of *time* is only applicable to the laws of nature so that events or measurements taken will be the same and thus be able to arrive at a more conclusive result. Such a time will only change when there is a change of event. Nevertheless, since the entire Universe is constantly in motion, we can safely conclude that every event can change with respect to time.

The universe is like a rare clock, such***that the engine being once set moving, all things proceed according to the artificer's first design, and the motions***do not require the peculiar interposing of the artificer, or any intelligent agent *employed by him, but perform their functions***by virtue of the general and primitive contrivance of the whole engine.*

—Robert Boyle

Now that we have finally addressed the major roadblock of the time factor in modern physics, we shall continue to pursue the kinematics behind the rotational motion of a semi-rigid system. Going back to our lighthouse experiment, we shall focus our attention to the behavior of light in a rotary system. Remember that we were still on the subject of a space-based observer being at a distance of twice the speed that light travels in a second; that is the distance 6×10^8 meters. Let us simplify this unit by calling it a light-second. A light-second is the distance that the light needs to travel in one second, which is 3×10^8 meters. So, in other words, the spaceman is two light-seconds away from the lighthouse. Assuming the lighthouse makes one revolution per second, the spaceman will see the signal for every two seconds on his watch. Please note that the difference between a light that is *directly* emitted by a source and the light that is received by an observer after the beam of that light is rotated is entirely different. It is therefore the objective of this experiment to identify the differences between these forms and present them to the reader in the most comprehensive manner.

The first difference lies in the measurement of time and distance. If the light were emitted in a constant flow, then the spaceman could never tell that how long it took the light to reach him (without the assistance of other measuring instruments) since he does not know his distance in space. In our experiment, instead of having the light emitted uniformly, we chose to have the light emitted at different intervals or time-pulses. In such a way, the pulses become a form of clock. Next, we used the rotating beam of lighthouse in order to have the light seen by the spaceman after a constant revolution made by the source. The end result therefore determines whether light or any form of semi-rigid body bends (and is scattered) in a rotary system.

Imagine several observers standing equidistant from a light source that emits uniform light in all directions. Let's assume that all observers will see the same light-time as long as they are within the same radius. If the light is emitting uniform flashes at a constant time interval and the observers are both within the same radius around the source, they will see the flash at the same time interval.

Let us now put the source to the test by making it rotate and imagine the results. As you can imagine, the first experiment shows that all the observers still see the same flashes at any position within the same radius. Once the source begins to rotate a little bit faster, they will start to see flashes in shorter time intervals. This happens not because the flashing occurs more rapidly, but because the rotations deliver the flashing in higher frequency time intervals.

Essentially, if we continue to increase the speed of the rotation, there will be a point of time when the flashes become a continue flow of light and this can be seen from any point within the same radius. This is the other difference. *A flash can be seen as a continuous flow of light under the influence of a rotary system.*

With this idea in mind, we continue with our lighthouse experiment and see whether it makes sense. Imagine the earth-based and the space-based observers, both at distances of one light-second from the source. We shall start rotating the light beam by one revolution per second. Without changing the locations of our observers, we will increase the speed to two revs. per sec. This time they will see two signals per sec or one signal per half a second. It is easy to conclude from here that if we continue to increase the rotational speed of the source, then the observers see more signals per second until, at a certain speed, the light source becomes a continuous flow of light.

Suppose we shift our space-based observer to a distance of two light-seconds from the source. Will the spaceman still see the signals as a continuous flow of light? In this case, no. Once he moves further out, the flow breaks into individual signals again. This is because the increased distance results in an increase of time taken for each flash to reach the spaceman. The tangential velocity of the first signal *would have to be larger than* the second signal in order for it to arrive at the same time given the increase of the radius when the angular momentum remains the same.

Skeptics will probably argue that since the light was emitted in straight and regular fashion, when did angular momentum come into the picture? Remember that during the initial stage the rotation of the source increased until the flashes simply filled up every space of the time intervals. That being the case, then every emitted flash will have a component of an angular momentum at the point of emission.

If this still doesn't convince you, think of it in another way. Suppose we trigger the source to emit light at very short intervals (within small fractions of a second between each flash) such that the light is seen to be a constant flow even without revolving at high speed. After which, we make the source rotate in the same manner described earlier and imagine what happens? If we assume that the angular momentum of the light to remain constant, then we will still arrive at the same deduction that the light signal emitted first will be the farthest and will therefore have the lowest tangential velocity. This is because the signal loses its tangential velocity proportionate to distance, but the linear velocity remains the same. The end result will be that the locus of the signals will be in the form of a curve. If this were to happen, then the signal as measured by the spaceman would be longer than clock that is given by the source (rev./sec). This lag of time will therefore give the spaceman the impression that the distance of the source is further than he thought. Further more, the trajectory of the light will have a deflected angle due to the losing of tangential velocity and therefore also create another impression that the source is changing its position.

10. The Third State of Rotation

The third state of rotations occurs when a fluid body rotates. The tangential velocity of a point is inversely proportional to the distance from the center such that they tend to act towards the center of mass.

Fluid (Free State)

Substances can be broken down to four states of matter known as solid, liquid, gas or plasma. There is a very fine line in trying to differentiate them. Unlike solids that tend to have volume and shape, liquids tend to keep their volume but not their shape, while gases will keep neither their shape nor their volume. The fourth state of matter, plasma, gets its form when gases are treated at very high temperatures, such that the electrons in an atom are stripped off into positive nuclei.

Hydrodynamics, the study of fluids in motion, is a fascinating field. Fluids are either liquids or gases, but never keep their shape and may vary their forms due to temperature, pressure or even density. The main property that distinguishes a fluid from a solid is that a fluid is unable to maintain a shear stress for any length of time. The behavior of fluids, particularly water, is very unpredictable and the flow can be in the form of ripples or waves, streamlines or vorticity.

Streamlines are lines drawn at a tangent to the velocity of the fluid to indicate the flow, rather like the lines we draw to describe a magnetic field. A steady flow means that the fluid maintains a constant velocity at any point of the flow. It is therefore easy to imagine that the flow of the fluid is quickly and constantly replaced by another new fluid moving exactly in the same way. Otherwise, in unsteady flow the streamline pattern changes in time and its pattern at any instant does not represent the path of a fluid particle. Often we describe a fluid in this state as experiencing turbulence.

Vorticity describes the flow of a fluid in such a manner that it behaves like the pattern in an Archimedean spiral. This is the familiar whirlpool we see in bathtubs. The reason why fluid can flow this way, unlike in rigid or semi-rigid materials, is that the properties of a fluid do not impose any restrictions (due to weak molecular force) on the motion of the particles. Thus, particles nearer to the center of rotation will have the freedom to travel *at a higher velocity* than particles that are further from it.

Without any limitations to the above, anti-collision behavior does not apply here and therefore, the molecular interactions are highly volatile. Recognizing this fact, molecules will collide and influence other particles to travel in a circular motion. Consequently, the entire fluid will evolve into a non-uniform circular motion. On a broader scale, molecules, though weak in molecular forces, will tend to influence (attract) one another into following the trajectory of a spiral. Consequently, the particles having higher energy levels will experience an increase in mass and thus, the particles are denser at the source.

We can illustrate this point by swirling water in a pail. If you apply your force to the middle of the pail (center of mass), you'll observe that no matter how hard you stir; it is very hard to create a huge spin—that is, a spin which drives the water all the way to the side of the pail. There are various reasons for this. Firstly, the moment of inertia is *larger* at the side of the pail versus the *small torque* in the middle. Secondly, the molecular forces between the particles are weak and so molecules are allowed to move freely and faster without having to obey the anti-collision principle. As a result, they tend to move freely without influencing each other greatly in their own restricted frame. Thirdly, there is a competition between gravity and centrifugal force here, in which the latter is trying to push the water out and upwards. The water moves upwards because the pressure from the side and the bottom of the pail prevents the water from flowing out. Essentially, the tug-of–war is always there when an object—be it rigid, semi-rigid or fluid—experiences rotational motion and the resultant force is dependent on the rotational properties of that object in rotation and the applied torque.

Now we'll change the set-up to make things easier. This time begin by stirring the water from the outside. When the water gains angular momentum, then we stir it from the inside. In this case you'll find that it's much easier and little torque is needed to produce a powerful spin! In fact, the spin could be so powerful that the core of water in the spin forms a "hole" or "whirlpool" in the center. This happens because the torque is greater than the gravitational pull on the water and eventually converts to centrifugal force.

Many conclusions can be made from this simple experiment. Firstly, you will see turbulence during the initial stirring. This is because water can flow very freely and therefore the molecules are not in uniform motion. Secondly, after the water has gained angular momentum, you can create a vortex by stirring in the middle. This vortex is easily attained because the momentum of spin requires only a low torque, due to the small radius and high angular velocity of the system. Note that a vortex only forms in the center of rotation, as though the fluid's center of mass is concentrated there.

Once a vortex has been created, the waves of the water do not behave in the manner of ripples anymore but in the form of spiral. Small floating objects dropped into the water, at the same time at different radii from the center will move at different rates. Those dropped outermost will move more slowly than the ones nearer the center (in contrast to the rotation of a rigid body where the speed of a particle increases proportionally with their distance from the center). Secondly, the objects nearer to the vortex get sucked in and the one that gets pulled in first will remain in the center while the next in line will "orbit" around it. The outer ones that don't get in will remain in "orbit" without getting sucked in at all. This phenomenon is similar to the behavior of planets and other objects in our solar system. However, neglecting viscosity in the system, random motions are experienced by the objects floating outside the vortex. These include one object moving faster than another due to higher angular momentum without collision. As well, some objects would travel in an elliptical path even though the pail was round.

11. The Fourth State of Rotation

The fourth state of rotation applies to any of the earlier three states with translation in two dimensions. The point in the center will translate with respect to a reference system while the rest, having a certain radius from the center, will describe the function of sine or cosine curve. In the absence of internal force, the point will have an angular motion about its own center unless the radius of the point itself approaches zero.

Rolling System

Now that we have an understanding of the kinematics of particles in rotary motion, it would be still more interesting to examine the behavior of such a system when experiencing both rotational and translation motion. A body acting in such a way can be said to be rolling. For simplicity's sake, let's assume that such a body does not slide when it rolls.

It is generally thought that a body in an isolated system will roll in a similar way to a body rotating about a fixed axis. In this respect, it is a common misconception that the particles in that system will behave no differently from those in other systems. As we have pointed out in the earlier chapters, a body in a different state of matter will give rise to different phenomena and hence different energies within the same system when acting under the same principles of rotation.

To begin with, let's examine the kinematics of a rigid body and make some observations. For simplicity's sake, we will assume that the body is a sphere does not change its shape or volume when it rolls. We will also assume that the sphere is rolling in a straight line (the plane of rotation is parallel to the direction of the motion). From here we may look at the sphere in a circular cross-section and identify three particles each having a different radius from the axis of rotation. The first particle 0 will be on the axis of rotation, particle 2 will be on the circumference of the sphere and particle 1 will be in between particle 2 and 0. Initially, particle 2 will start from the point of contact with the surface.

When the body starts to roll, three of the identified particles will behave differently during motion. Take for instance particle 0. It will undergo translation with spinning (unless the radius of the particle 0 reaches zero) and still maintain a fixed orientation throughout the motion. Fixed orientation because its distance is always the same from the point of contact and the axis of rotation. Its motion is therefore parallel to the direction of the body that it translate and every change in distance of the point of contact P is equal to the distance changed when it translate linearly. In this respect, it can be seen easily that this particle 0 not only has an angular kinetic energy about its own axis but also the linear kinetic energy with respect to an external reference system (the surface in this case) and its always directed towards the direction of the motion of the body.

Let's look at particle 2 now and note the extreme contrast between its motion and that of particle 0. First, we start particle 2 at its point of contact on the surface. From here, every change of distance that particle 0 translates will result in a change of particle 2 with respect to particle 0. Therefore it is safe to assume that both of the particles traveled with a same distance within the same time interval. This contrasts with the case of a body rotating about a fixed axis without translating.

Remember in our discussion of the rotation of a rigid body in chapter 8 our particle 2 will move a certain distance while particle 0 being the axis of rotation will remain stationary (assuming that particle 0 is fixed or spins in a negligible way). During rolling, particle 2 will maintain a fixed radius with respect to 0 while changes the distance with respect to point of contact P. Interestingly, particle 2 will accelerate much faster than particle 0, achieving twice the velocity when at the top of the rim with respect to the point of contact P. At this stage, the kinetic energy of this particle is said to be the highest. After this stage, the tangential kinetic energy will drop as the particle decelerates when it reaches the surface and until it stays instantaneously at rest. The curve that is traced out by particle 2 looks like a "dumb-bell" and is known as a cycloid.

This phenomenon contrasts greatly with the motion of a rotational body. The particles in a rotational body will continue to move in a uniform circular motion, providing a constant angular kinetic energy. In contrast, the tangential kinetic energy of a rolling body changes with respect to **P** while maintaining a constant angular kinetic energy with **0**. From this, we can deduce that particle 2 experiences two types of kinetic energy: first, the angular kinetic energy with respect to **0** and second, the tangential kinetic energy with respect to **P**. The position variable of particle 2 makes the whole system dynamic. The angular kinetic energy will remain constant during rotation while the tangential kinetic energy will change throughout the motion. As such being the case, the total energy of a rolling system will therefore have to be the sum of the rotational kinetic energy and the tangential kinetic energy or potential energy with respect to **P**, with the latter being dependent at the point in space. As a result, the total energy of the system is conserved. The rotational kinetic energy is conserved by the law of conservation of angular momentum while the tangential kinetic energy is conserved by the principle of the simple harmonic motion; which tends to oscillate from potential energy to kinetic energy

Next let's examine the movement of particle 1. Particle 1's behavior is similar to that of particle 2. The difference is that, like the case of a rigid body in rotation, particle 1 will have to travel slower than 2 but faster than 0. However, there is more. Particle 1 will have a motion similar to that described by simple harmonic motion. Simple harmonic motion means that a movement repeats itself in a regular and predictable way. It is also known as rhythmic or periodic motion. Our heartbeats, the cycles of seasons or the orbital motions of planets and electrons are all examples of rhythmic motion. This motion is capable of restoring forces and its position of a point varies with time as a sine or cosine curve. It is important to us because we can see it in most laws of physics. Particles experiencing SHM are able to conserve and restore forces, just like when a pendulum conserves its total energy, even when it oscillates from potential to kinetic and back again to potential energy again. This allows the forces of nature to act in dimensionless and infinite ways. In another words, this is another demonstration of the principles of rotation.

In view of the above, it is therefore easy to see that the particles in a rolling body exhibit kinematics with particle and wave properties. For example, if you look at particle 0 you may say that it propagates in a linear and angular motion. If you look at particle 2, you will say that it moves with a function of a transverse with a chaotic effect. Particle 1 will propagate in a sinusoidal motion, that is a motion that follows the curve of a sine or cosine along with time. It is no surprise then that a rolling system can manifest different behavior when in motion. It is therefore necessary that we must not limit our study and application of a rolling system to a rigid body. Correspondingly, our analogy to a rolling system can be further applied to a semi-rigid or even fluid body and still able to yield different results.

So far, we have only considered an isolated rolling system in a straight line; that is a system in the absence of external influences or external forces. Let us continue to examine another model—the Sun and the Earth. First of all, the Earth does not orbit in a straight line. Second, the Earth acts under the law of universal gravitation. When the Earth orbits, the center of the Earth will maintain a fixed orientation towards the Sun and the imaginary point of contact will be such that of distance on Earth is the nearest towards the Sun. However, the center of the Earth will not be moving in a linear motion but in a circular motion with respect to the Sun. If this is the case, then we can say that the center has a constant angular kinetic energy and the angular momentum is conserved under the law of conservation.

What constantly changes are the particles other than the ones on the axis of rotation that we discussed earlier. For example, if we fix an arbitrary point on the surface of the Earth and examine the dynamics much like the case of

particle 2, we can conclude that the point 2 would experience higher kinetic energy at the point when it is furthest
away from the Sun and the gravitational pull from the Sun
would have been the least at that instant. Contrariwise,
another point on the other end would have a same rotational kinetic energy and a higher tangential potential
energy with respect to the Sun. The gravitational pull is
then stronger as compared to the former. Hence, the total
energy is conserved between the two arbitrary points. The
center of mass will orbit in an angular motion with an
acceleration directed towards the Sun but the kinetic
energy will continue to act tangentially and parallel to the
direction of the orbit.

Next, let us choose another arbitrary point in between the
surface and the center of the Earth. Suppose that this point
is fixed and rigid from the axis of rotation, then we can see
that the result will be the same as that of particle 1; except
that the trough and crest of the motion will be affected by
the gravitational pull of the Sun. In this case, the total energy
of *that point* in Earth will be dependent on its distance with
respect to the Sun. However, we must not forget that we only
assumed that that point is having a fixed radius with respect
to the center of the Earth. Imagine now that the point is not
fixed but of a similar character of a semi-rigid or a fluid
body? It would then be logical to think that the energy of the
fluid particles in any part of the Earth would be greater than
the surface. This may not be necessarily true as the object

here is constrained by the analogy of a restricted frame discussed earlier. In other words, if an object has a character of a rigid cum fluid property, then the energy *in* the object can only *be less than or equal* to the energy on the surface provided that everything is to remain unchanged.

12. Curved Geometry

God draws straight lines as a point on a curve.

The Euclidean geometry that we were taught in school is somewhat different in what we are going to learn about warped space-time. The shortest distance between two points is a straight line and parallel lines never meet. This view works very well on a flat piece of paper, that is, a flat geometry. In a curved geometry, the shortest distance is not only a curve; it is also the one that is nearest to the origin of the curve.

What is a curved geometry? Is it a circle or is it just a line that is curved like the arc of a circle? The answer to this question is both yes and no. Yes, because it is true that a circle or arc is part of the geometry and no, because it is more than these. You have probably heard something about curved space-time and that the presence of the heavenly bodies can cause space to curve and therefore the path of the other planetary systems to follow the same so much so that even light (which travels at the speed of 3×10^8m/s) has to conform to this path. There is so much hype about this that Einstein has become a living legend till this date.

So how important is this idea to us? We see it in the wheels of the car or the gears in the machine that help us to get work done. The circles and spheres that we learned about in school were merely useful parts of our mathematical grades. Aren't we glad that God has designed the earth to be round so that we don't fall off the edge of the world and that it rotates so that we have a day's time to work and a night's time to play and rest. Best of all, we have the wonders of the four seasons to appreciate the beauty of nature. It will be an endless pursuit if I were to continue to explain the laws of nature that were prescribed by God. Nevertheless, we should recognize this fact and give credit to the Creator. In fact, we have often admired the works of the engineers when have actually exploited and leveraged on the knowledge that was engineered by God. If God were to file a patent for his invention, every one of us would get sued!

Let us start by exploring the earth. Assume that both of us are total strangers on earth and we have only met once, shook hands and promised to go on our own separate ways and discover our own world. So we part, with me going east and you going west along the equator. (Let's assume we started from Singapore and can walk upon the waves.) Days past, weeks lapse, and we just kept going. Then one fine day, we met again and become bewildered by the coincidence. We then argue that one of us must have broken the promise and were determined this time never to meet again. This time instead of going east, I go north and you where you are.

So I journey northwards, cross over the pole, and keep on going along the curve of the earth, south and then north again, and eventually I'm back in the place where I began, facing you. No matter how we design our route, we are destined to meet again and be friends. Later, we have a chance (let's use our scientific imaginations here) to look at a piece of rock that resembles much like the shape of a sphere and begin to study it. We begin to realize that the shape of the world is the same as the shape of the rock. So this is the reason why, by the properties of the shape of the sphere, we always meet.

Let us elaborate this point with a bit more clarity. Suppose that we live in a world that is confined in the space of a balloon. Suppose also that there is another person living on the surface of the balloon (edge of the Universe if it is to be finite). After making some measurements, we decide to pay him a visit and embark on a journey at the speed of the light. During the midst of the journey, we realize that our target is somewhat off for reasons that we don't really know. All that we notice was that our friend seems to be getting further away from us (by the signal of the light that he has given to us) and his position seems to be ever changing (by the direction of the light that we received since we can't use a compass) and worst still, our space ship has to constantly steer to adapt to the flow of the geometry in space such that we don't get entwined into another gravitational wave (sometimes it helps) in order to be able to achieve the shortest possible route to meet our dear friend for Christmas. This error is probably due to the fact that the measurements taken were meant for a straight line. Assumptions made were therefore invalid because the balloon (space) is curved and as the balloon gets inflated (space expanding) the distance just gets further. Furthermore, due to optical illusions, the direction of the light that we used to guide us also gave us the wrong route (gravitational lensing). The example will be very much similar if we are on the surface of the balloon. When the balloon gets inflated, the curve that is between us just gets longer. Such complexities actually happen in the Universe.

So much for the works of nature. Let us now continue to pursue the subject further. You may have pondered why most things are round in one way or another, be it their shape or their motions. Is it really merely coincidental? I reckon not. Roundness has evolved as the fundamental geometry as it is the most practical shape that conforms to laws of nature. You can contest this by examining numerous elements of the universe, from a microscopic to a macroscopic scale. You will doubtless arrive at the same conclusion–most objects are curved.

This is the reason why curves have been treated with such significance right from the beginning of this book. Every detail has been carefully thought of and explored with optimal clarity so as to highlight this significance to the reader, omitting the mathematical expressions which essentially is only useful to scientists who need the quantify the concept in the terms of a value. Without limiting ourselves to just particles in a circular motion, we shall now further explore the behavior of an object in circular motion in curve geometry.

If you drive a car and need to get to a destination within the shortest possible time; there are two things you can do. One, you drive at the highest possible speed (assuming there is no speed limit) or you take the shortest distance and the correct direction or maybe even all of the above. Technically speaking, in order to reach from point A to point B; time, distance, speed and direction are correlated. If all of these can be achieved at the optimum level, then it can be understood that even without the derivations of mathematical formula, any layperson will know that *it is* the shortest possible route. Note that *route* is used in this case to describe the locus of the motion.

In school, we are taught that the shortest distance between two points is a straight line. This is undoubtedly true but still it needs some rephrasing; the shortest *possible route* between two points is *probably* a straight line with speed, time and direction being equal. In other words, it could be an arc within a square, a u-turn inside of rectangular turn, or perhaps a circle within a circle.

FREQUENTLY ASKED QUESTIONS

Reader: all these ideas are so ambiguous. Why can't we just move from point A to point B? Why do we have to travel in other forms and so forth?

Physicist: The answer is not so much *why* as *how to*. Our Universe was not created in a geometry described by Euclid. It was created by nature's geometry such that all physical laws of motion conform to it. That is to say, in order for the entire Universe to work under a simple system and yet intelligent enough to govern all the heavenly bodies, the set of rules have to be fixed and yet versatile enough such that all the bodies behave in a same manner whereby all forces on the motion are equal and constant. Hence, if all the planetary systems were to travel in a linear motion or some were to travel straight and some in angular motion, what then would happen to this world?

I am trying to point out here that we can't travel straight because the world that we live in doesn't allow us to do so. It has so many structures (e.g. Buildings, parks, shopping centers, etc.) that constrain our movements. Likewise, such is the characteristic in a system, which is being deployed in the entire Universe. Therefore, in order to arrive within the shortest possible time, *the shortest possible route is to travel in a direction towards the object such that the locus (definable mathematically) has the least distance that is allowed in a system.* For example, if you want to go to the office from your home you will need to travel a certain distance in a given time. If you tell me that the distance of your office from your home is 70 km and the time taken is an hour and if you drive with the speed of 70 km/h, theoretically, you should reach your office at the stipulated time. However, there were times when you were early and perhaps, there were times when you were late (assuming that you leave your home at the same time each day).

Reader: Why is this so?

Physicist: traffic conditions, weather conditions and so forth (refer as medium of path).

Reader: when I measure the distance between your house and your office using a straight-line rule, the distance is only 50 km and so you should reach your office earlier than the stipulated time, since by virtue of your speed at 70 km/h.

Physicist: I can't because there is no straight road that links right to my office and so I have to make turns and bends, which ended up with a longer distance (refer as path of geometry).

Reader; if you travel at twice the speed, then the time taken should be less?

Physicist: yes, the faster you travel the lesser the time you will require. In other words, time is not an absolute, it is a variant. If you travel at the speed of light to your office, the time taken for you to reach your office could be instantaneous such that when you look at the watch, the time is the same as the time that you have left your house, meaning there was no change in time! In reality, we are not allowed to do so because we cannot travel at the speed of light. It is for this reason that classical physics has been valid for such a long time. Measurements (observations) made within the classical mechanics were in a predictable sense because it was not traveling at the speed of light. However, *in a rigid rotating system*, the time taken for any two particles with the same angular velocity is the same at any point of the motion.

Reader: so which is the correct distance 70 km or 50 km.(refer as real distance and ideal distance).

Physicist: it all depends on your definition of distance. If you are referring to the distance that is on the map, then 50 km is the distance. If you are referring to the distance that you need to travel, then 70 km is the distance. Scientists usually use the latter to make measurements because that is the real distance that a body or an object needs to travel. However, even with such knowledge, it is still very hard to quantify since bodies that are in motion are perpetually and relatively in motion. On a universal scale, the real distance now could be the real distance sometime ago.

Take for example the case of the Earth around the Sun. There is no doubt that the earth moves in an elliptical path around the sun. So would it be easier to travel to the sun when the earth is nearer or further? The answer is quite obvious; it has to be when the earth is at the nearest point in the elliptical path. Likewise for the rest of the stars that is constantly moving, if the Universe is forever expanding and the stars are perpetually moving away from us. For that being the case, then how could we reach them or see them (light coming towards us)? How are we sure that what we see today is what we should really see? The factors that disturb this fact should then be the other variants that constitutes to the resultant route of the light. Recognizing this fact that even light has to conform to the geometry of the Universe that God has prescribed, light when traveling in rotary geometry, will translate in a motion designed by that geometry. A classical example of the behavior of light when passing through a medium with a different density, say a convex or a concave lens can yield different results.

Reader: is your direction correct?

Physicist: this is a little bit hard to answer. It all depends on which road you are taking. Some roads can only traveled in a certain direction so the road that you driving to your office and the road that you are driving back can be different and hence, the distance traveled can be different.

Case no.1; distance traveled to office is 50 km, distance traveled back is 40 km. So what is the real distance? The answer lies in the reference frame again.

Case no.2; paradox of the merry-go-round. Tom and I were sitting on the merry-go-round and you were standing outside. We were rotated with one revolution and we arrived back at the same place at the same time. The distances traveled were different and yet the destination was the same. It was as though we have never traveled because Tom and I met you again at the same place! But Paul knew that we did move and he saw that I was moving so much faster than Tom.

Case no.3; this one requires a little bit of imagination. If a distant star is measured to be of x million light-years from the earth, would it still be the same x million light-years if we were to travel towards the star and the star is also heading towards us at the same time?

If we attempt to define the shortest possible route as the shortest possible time within the shortest distance in the same direction then *it should be the spin in the center of the axis.*

In General Relativity, gravity designs the geometry of the Universe. The curvature is caused by the presence of mass. Whilst this is true, we mustn't forget that all heavenly bodies are in motions and massive moving bodies can create *influence* far greater than one can imagine. As a matter of fact, the Universe should not only be curved but also spiraled. This *gravitational influence* can be so great that I am beginning to feel that the Universe that we see today is distorted to some degree.

The *gravitational influence* that I am relating is in the form of a spiral and that even light has to escape under the geometry of this spiral. The fall of big object versus a small object is the same simply because the influence is the same so long they are in the same frame of reference.

13. Law of Gravitation

Gravity and acceleration are indistinguishable.
—Einstein

Gravity is the most common of the four forces and the weakest among them all. We hear of gravity in almost all textbooks and physics classes. It has the longest effective range and is also positively attractive. Gravity has the greatest influence over our daily life. It is one of the laws of nature that govern the dynamics of the Universe. Gravity is the collective force of mass of the particles within the body. This collective force is further exacerbated by the rotational effect, which extends throughout the area of influence and stops at the line of break. After which, the spiral effect takes control. As we've already discussed, the inverse-square law states that gravitational force decreases as objects become more distant. As an object moves further, it loses its tangential velocity with respect to the center and, therefore, objects take up the form of a spiral. Another way to conceptualize this is if you hold a string and swing it about your hand, you will notice that as the length of string is increased, the end of the string gets "slower" as it loses its tangential velocity. Neglecting air resistance, the entire string becomes curved, and if geometry allows, then the string will take on the form of a spiral. This is actually an application of the second state of rotation on a semi-rigid body, and it is this that designs the geometry of the Universe. Just look at the shape of galaxies, for instance.

The gravitational pull of a massive object does not only depend on its mass, but is also dependent on the speed of the rotation. A body like planet Earth can be seen rotating with a pull that works from outside to the core. On the surface of the Earth, a force is directed towards the center of the Earth due to attraction of the trillions of particles that made up the Earth. These particles have attractions of their own which, collectively, forms a large *gravitational influence.*

Let us examine some of the secrets behind gravitational force that have fascinated humans for centuries. We all know that the earth is made up of a huge mass of particles and each of these particles has a mass-energy of their own too. First, we shall focus our attention by looking at two particles like the one we have done for the rotation in the rigid body.

If we rotate such an object in this state, the resultant motion of any of the two particles in the frame has to be inward. Two reasons can be derived from this. Particle 1 can only travel in a rotational motion (due to the anti-collision analogy in a restricted frame). Second, the change in velocity in 1 creates an acceleration directed towards the center.

As a result of this, on a collective scale, it is therefore easy to see that all particles in the same restricted frame act towards the center, which is the center of mass. It is only the particles on the surface of the earth will experience a centrifugal force, perpendicular but not equal to the centripetal force. Having said this, it should be clear that the centripetal force is larger than the centrifugal force since the sum of the mass inside the earth is very much greater than the sum of the mass on the surface of the earth. As a result, the total angular momentum is larger nearer to the surface of the Earth, typically near to the equator.

The Sun and Earth (or any other planets) obey the universal law of gravitation. This law simply states that mutual attraction of two bodies mass M and mass m is inversely proportional to their distance squared. Imagine a string which links the centers of the sun and the earth. The earth with its presumably lower mass, will only be allowed to travel in a orbit much like the case of a satellite orbiting around the earth. But unlike the case of a rigid body, the earth is allowed to move in a path formulated by Kepler's laws of planetary motions and yet confined only by the space and the gravitational pull of the sun. As a result, the locus of the orbit becomes elliptical. The counter-clockwise motion of the earth's rotation can also be explained by imagining that the line is tied to the center of the earth with the tangential force on the earth moving in a linear motion.

While the earth orbits around the sun, it produces an angular kinetic energy about its own axis as the orientation and position changes relative to the sun (see paradox of merry-go-round). The resultant effect of the earth rotating in a counter-clockwise motion is very much the same as a car's wheel rotating on a road while translating a linear motion. If this wheel is to move around the equator of the earth, an observer in space will deduce that the wheel is acting under the influence of the gravity while trying to translate a linear motion. In other words, the spaceman will see that wheel is orbiting around the Earth. We can also imagine and draw a "chain drive" like the model that we see at work in the gears in a bicycle Here is an example of Sun and the Earth, and another is the wheel moving on Earth.

Conclusions

Star Wars

Perhaps particles are essentially "boxes" that hold energy. Indeed, modern views of the origin of the mass of elementary particles have very much this character. Perhaps there is no meaning to the concept of mass separate from energy!

—Robert Adair

We have long known that stars and planets, from the beginning of the universe, have formed with the help of gravity. Still, we cannot help reverting to the same ageless questions: how did gravity itself come about, and what caused gravity to coalesce the gas and dust widely diffused across the early universe?

In order for us to understand this natural phenomenon, we need to reverse time's passage by looking back to the beginning of the universe. If we assume that the universe started with an extremely fine scattering of diffuse gas and dust across empty space, then a mystery arises. Since the law of inertia states that any body without motion continue to stay at rest, it doesn't make sense that diffuse particles suspended in empty space should form into masses. By all rights, nothing should have happened.

The most widely accepted theory about the origin of the universe is known popularly as the "Big Bang", an event that happened some 15 billion years ago. This historical event involves a massive, violent explosion that sent energy and matter through every directions across space and time. In this respect, space and time were created after the Big Bang. Logically speaking, time could not exist before the existence of the universe, nor is there any space outside the universe. Of course, the fundamental question now is: what triggered the Big Bang?

Before we get ourselves entwined with a string of questions, let us examine the proof of this theory. Based on observations by astronomers, there is ample evidence that the galaxies throughout the universe are receding from one another. If we turn back time, it is reasonable to suppose that the galaxies would have started from the same space and at the same moment. This supports the idea that they are all moving outwards from a single, explosive origin point. Another piece of evidence comes from the detection of the "cosmic background radiation", first witnessed in 1964. The temperature of this radiation leftover from the initial explosion has cooled down to about -270°C, well in accordance with predictions of the Big Bang theory. Along with some other observational results, the Big Bang theory would seem to be indisputable.

For the sake of our inquisitive minds, we might try and guess as to what triggered the explosion. Let us simplify our query by not asking when it actually happened but rather what caused the event. In this case, we can start by imagining that we have a certain volume of gas contained in a certain space. Though loosely scattered, these particles have charges of their own due to their own angular spins. If one particle happens to contact another with like charges, this will cause them to repel each other and hence, erratic motion occurred. If a particle happens to move near one with unlike charges, then they get attracted to one another. The attraction alone will cause a shift in the center of mass and produce an angular momentum by virtue of their angular spin. In both cases, motion begins. This self-reinforcing process continues to attract and repel particles and thus provides a constant angular momentum. As time goes, the material gets denser and denser.

While the angular momentum remains constant, the overall spinning motion accelerates the particles inwards at higher and higher speeds. This is because as the distance to the center decreases, the tangential velocity of the particle with respect to the center of the system and the angular velocity about the same particle's own center continues to increase, as described in the third state of rotations. When this happens, the accelerated motion causes the particles to collide at very high speed with an infinite energy at the center, squashing and crushing the particles out of recognizable existence. Eventually, the lightning collision of atoms is so violent that their electrons are stripped off, leaving a gas made of bare nuclei and electrons. At this stage, nuclear fusion begins—such that the nuclei are forced to stick together. This reaction will produce huge amount of heat and create an enormous amount of pressure in the center. However, at this point of time, the particles still have nowhere to escape and the energies are still trapped because new matter continues to be added at the circumference. Therefore, the swirling mass continues to suck every existing particle, providing a constant fuel for the nuclear reaction to happen and waiting for the catalyzing moment to occur.

By this time, all particles in the system would have been engaged in the spinning and is trapped in a ball of fire, very fiery but not powerful enough to overcome the angular acceleration towards the center. Though condensed, there is still room for further collapse. This is because the clump of matter is still in the shape of the ball. The rapid spinning starts to exert a pull on the outer regions of the top and bottom of the ball, causing it to be compressed and flattened (as described by the first state of rotations—the tangential velocity increases with the distance), compressing the matter even further. By now, temperature would have reached millions of degrees centigrade. Soon, the pressure would propel every participating particle to move at lightning speed. Hence, the tangential velocity of each particle will start to increase with the distance from the center (a reverse order from the initial stage). By now, particles having the greatest circumference would have reached the highest possible speed, too much for it to be directed toward the center anymore. The action refuses to stop here. While the particles at the furthest reaches the maximum speed, the tangential velocity of the particles that are in between the center and the furthest continue to increase, testing the limits of the laws of nature. By this time, the particles fueling the explosion have been depleted; the pressure is no longer trapped. When this happens, the extreme heat in the core continues to build up intensely, causes matter to ignite for the final time. This time with a bang!

The resultant explosion sends every piece of matter spiraling and expanding simultaneously. Particles having the furthest distance get discharged first. Losing some mass, the spin of the central mass remains so fast that it starts another self-reinforcing chain reaction (as described by the second state of rotations). Every piece of matter will be blown away with tangential motion about the center of the system and an angular moment about its own center, all having the same direction. From here, matter starts to cool down, the massive pieces will form their own system known as a galaxy and influencing the smaller ones into orbits around it. Some of them will continue to burn with the fuel left from the explosion and the denser ones would have mini-explosions of their own. Our sun is one of them, self-sustaining from the hydrogen atoms collected from the initial explosion and swirling the rest of the debris (our planets) into orbits around it.

Not every system is fortunate to have a star like our sun. The destiny of a star depends very much on the mass it was given. Firstly, it needs to have at least 10 per cent of our solar mass in order to generate nuclear fusion. If it has, then the supply of the hydrogen gas could last for billions of years. But once depleted, the pressure generated from the heat will disappear, causing the core to collapse due to the spinning again. Once again, heat is created but this time without any nuclear fusion. This form of heating will cause the star to radiate and expand beyond the original size, swallowing and vaporizing other matter around it. What remains at the core are densely packed nuclei and electrons, and the collapse stops here. The star will still glow for a while before it starts to dim like a dying ember. The characteristic white-glow of these stars lends them their name of **white dwarfs.**

If the star is five times more massive than our sun, its death is even more drastic. As the mass is many times larger, the contraction is even stronger. It is so powerful that the central core turns into solid iron and this doesn't stop until even the iron fails to withstand the contraction. When the iron finally gives up, it will rip itself apart and explode into empty space. This is known as **supernova explosion.**

The story hasn't ended even at this stage. The remainder of the star after the supernova explosion continues its journey of its final phase of collapse. When the original star was 10 to 20 solar masses in size, the final collapse after the explosion can be so powerful that even the electrons are forced to merge with the nuclei forming a huge volume of neutrons known as **neutron star**. Along with the rapid spinning, the huge mass is crushed to within a few kilometers in diameter. Believe it or not, a tiny speck of this could weigh as much as an aircraft carrier on earth.

Besides the neutron star comes its more exotic and more famous cousin, known popularly a **black hole.** When a star with an original mass of more than 30 solar masses collapses to a single point, the gravitational influence is so great that not even light traveling near the vicinity of such an object can escape it. Therefore, it "appears" black or rather it cannot be seen. This being the case, you may ask how scientists are able confirm the existence of black holes? The only way to do this is to detect its gravitational influence on another visible star. When the visible star orbits around an unseen object, that object will suck in the gas atoms liberated by the visible star and tear the atoms apart. In the course of it, x-rays will be radiated and thus, be detected.

So it seems plausible that the entire universe's mechanical principles are based on rotations. General observations of satellites' orbit about the planets, planets orbiting about a central star, stars spiraling about a galaxy and finally galaxies spiraling out of the center of the universe demonstrate that rotational dynamics are at the base of all physical systems in the universe. Logically, it follows that the Big Bang was also triggered by a *rotary* explosion right at the beginning.

The first neutron star was not discovered till 1967 when an astronomer named Jocelyn Bell detected flashes of radio waves from a distant source. Initially thought to be some signals from an extraterrestrial being, it was soon confirmed that the flashes were given out by the star due to its rapid spinning—like the example of the *lighthouse experiment.* Later research revealed that such a star is capable of spinning about hundreds of times per second, so much so that a billion tonnes of matter is forcefully packed within each cubic centimeter!

Obviously, the dynamics of rotations have been able to provide a consistent life cycle of a star, from its birth till its death. Our way of seeing the world would be more comprehensive if we incorporated the idea of rotational mechanics into our view. The four states of rotations have clearly been able to manifest its ability to describe the dynamics of the entire universe, from the infinitesimal to the astronomical large bodies. Hence, the principles of rotation shall not be excluded from the physical laws of nature.